my DOG has a CRUCIATE LIGAMENT INJURY

– but lives life to the full!

A practical guide for owners

Hubble & Hattie

www.hubbleandhattie.com

Hubble & Hattie

The Hubble & Hattie imprint was launched in 2009 and is named in memory of two very special Westies owned by Veloce's proprietors.
Since the first book, many more have been added to the list, all with the same underlying objective: to be of real benefit to the species they cover, at the same time promoting compassion, understanding and co-operation between all animals (including human ones!)
Hubble & Hattie is the home of a range of books that cover all-things animal, produced to the same high quality of content and presentation as our motoring books, and offering the same great value for money.

More titles from Hubble & Hattie

Animal Grief: How animals mourn each other (Alderton)
Cat Speak (Rauth-Widmann)
Clever dog! Life lessons from the world's most successful animal (O'Meara)
Complete Dog Massage Manual, The – Gentle Dog Care (Robertson)
Dieting with my dog (Frezon)
Dog Cookies (Schöps)
Dog Games – stimulating play to entertain your dog and you (Blenski)
Dog Speak (Blenski)
Emergency First Aid for dogs (Bucksch)
Exercising your puppy: a gentle & natural approach – Gentle Dog Care (Robertson & Pope)
Fun and Games for Cats (Seidl)
Know Your Dog – The guide to a beautiful relationship (Birmelin)
My dog is blind – but lives life to the full! (Horsky)
My dog is deaf – but lives life to the full! (Willms)
My dog has hip dysplasia – but lives life to the full! (Häusler)
My dog has cruciate ligament injury – but lives life to the full! (Häusler)
Older Dog, Living with an – Gentle Dog Care (Alderton & Hall)
Smellorama – nose games for dogs (Theby)
Swim to recovery: canine hydrotherapy healing – Gentle Dog Care (Wong)
Waggy Tails & Wheelchairs (Epp)
Walking the dog: motorway walks for drivers & dogs (Rees)
Winston ... the dog who changed my life (Klute)
You and Your Border Terrier – The Essential Guide (Alderton)
You and Your Cockapoo – The Essential Guide (Alderton)

Acknowledgments

We are very grateful to Dr Anke Lotze, who gave us the impetus to write this book, and vet Dr Peter Morlock, who was always there for us with an encouraging word or two.
In particular, we would like to thank all our dogs: Dago, Kalypso, and Argos, and all the others with whom Barbara Friedrich has lived and worked, Dr Kirsten Häusler's dog Nala – gone too soon – who taught us many things, and, of course, her current dogs, Itchy and Dexter.

Whilst the authors and Veloce Publishing Ltd have designed this book to provide up-to-date information regarding the subject matter covered, readers should be aware that medical information is constantly evolving. The information in this book is not intended as a substitute for veterinary medical advice. Readers should consult their veterinary surgeon for specific instructions on the treatment and care of their dog. The authors and Veloce Publishing Ltd shall have neither liability nor responsibility with respect to any loss, damage, or injury caused, or alleged to be caused directly or indirectly, by the information contained within this book.

First published in English in July 2011 by Veloce Publishing Limited, Veloce House, Parkway Farm Business Park, Middle Farm Way, Poundbury, Dorchester, Dorset, DT1 3AR, England. Fax 01305 250479/e-mail info@hubbleandhattie.com/web www.hubbleandhattie.com
ISBN: 978-1-845843-83-0 UPC: 6-36847-04383-4 Original publication © 2011 Kynos Verlag Dr Dieter Fleig GmbH, www.kynos-verlag.de
Readers with ideas for books about animals, or animal-related topics, are invited to write to the editorial director of Veloce Publishing at the above address. British Library Cataloguing in Publication Data – A catalogue record for this book is available from the British Library. Typesetting, design and page make-up all by Veloce Publishing Ltd on Apple Mac. Printed in India by Imprint Digital.

Contents

Chapter 1

Cruciate ligament and cruciate ligament tear

Argos is a very lively 7-year-old Hovawart with a huge enthusiasm for both playing and 'working.'

Whilst out on one of our walks, I send him running up a big hill. After about five metres, he cries out suddenly, holding up his left hind leg, and whines so loudly that it makes me shudder. I run to him, expecting to see blood everywhere, but there is none. I stroke him, talk to him reassuringly, and when he has stopped whining, try to coax him into walking a few steps. He limps, and holds up his left paw – he is quite obviously in a lot of pain.

First aid

Slowly and gently, I stroke his right leg from the outside of the hip down to the foot, two or three times, rather than stroking the left

Argos, a 7-year-old Hovawart.

leg which is the one causing him pain.

So why stroke the pain-free side first?

Right from when Argos was a puppy, I have always examined both ears, both legs, both eyes, etc, regularly, in order to get him used to having a physical examination. It's a good idea to get your dog acquainted with this kind of contact, before an examination – for an inury, say – becomes necessary, because then he will be upset, extremely reluctant to co-operate, and afraid to be touched, in case it causes more pain. If he is familiar with being gently examined, it will seem normal to him and may even help to calm him.

You could play a game with your dog or puppy when he is fit and healthy, where you playfully investigate different parts of him such as 'belly!' 'nose!' 'ears!' etc. A positive side effect of this game is that it will strengthen the trust and bond between dog and owner, and ensure that he is used to being examined in case of an emergency situation.

Using the same method as for the unaffected leg, I gently stroke the painful left leg. I can't feel that anything is wrong and there isn't a foreign body between the paw pads. But with any slight movement of the left leg, Argos jerks back violently and emits a howl of pain.

Caution!

When investigating the reason behind your dog's discomfort, be very cautious. Go slowly and carefully, and only examine him if he seems fairly relaxed. Your dog loves you very much and will not want to hurt you – but even the best behaved dog in the world could snap if in pain.

I wondered if perhaps he had sprained his ankle? Or trodden on something sharp? Or maybe he had a sharp stone painfully wedged between his paw pads? But there was nothing like this.

What to do next

With much coaxing and treats, I lead Argos slowly, step-by-step to the car, and help him into it. He appears calmer now. We drive home, where he lies on the floor. After two hours, I let him get up – the rest doesn't seem to have relieved the pain, so I call the vet and describe to him what has happened. Luckily, we are able to get an appointment straight away.

At the surgery, three of us lift Argos onto the examination table, and while I stroke Argos' head, ruffle his ears, and talk softly to him, the vet examines his leg.

It quickly becomes apparent that Argos has a 'torn ligament.'

But what does this mean?

Chapter 2

Anatomy of the canine cruciate lilgament

People and dogs are similar in their physical make-up, but not identical. Unlike humans, dogs walk on their toes, and when they stand, their legs are not straight but bent at the knees. This means that their knees are subjected to greater stress than are ours.

The knee is the joint that connects the upper and lower leg. It is a hinge joint; ie it allows flexion and extension of the leg. In contrast, a ball joint allows rotation. The hip joint in humans, for example, is a ball joint, which allows us to move the leg forward, backward, and sideways.

Ligaments at the front and rear of the leg connect the thigh bone to the shin bone. Although this sounds relatively simple, anatomically speaking, it is quite complex to explain in layman's terms. Therefore, the following is a brief explanation of ligament function:

• the anterior cruciate ligament (ACL) keeps the tibia bone in the right position and prevents it from sliding forward
• the posterior cruciate ligament (PCL) stabilises the knee; ie prevents the leg from bending forward at the knee
• every bone in the joint area is covered with a layer of cartilage. The joint (hyaline) cartilage has the job of protecting the knees and

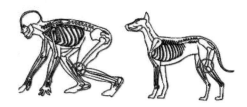

Positions of the human and canine knee.

bones by absorbing any stress. Hyaline cartilage is characterised mainly by its high-pressure elasticity

- in addition, another special part of the knee – the menisci – act as shock absorbers in a similar way to intervertebral discs
- the synovial fluid is produced by movement of the joint capsule, and provides the majority of nutrients to the structures in the joint, because there is almost no blood supply there

Maybe you know someone who has a torn ACL because of an external cause such as a sports injury? Previously, it was thought that, in the case of a dog, sudden, extreme weight-bearing was responsible for such an injury, but, as the ACL is very resilient, a huge force would have to be exerted on the knee to tear it – and weight-bearing of this degree is unlikely. Investigation of numerous ACL tears has allowed doctors to identify smaller, more gradual tears (perhaps caused by an awkward fall), which indicates a degenerative process, rather than a sudden tear.

Tearing the cruciate ligament can also cause other injuries, for

Case history
MAX

Max, an eight-year-old St Bernard, is a slightly tubby dog who adores his food. He also loves ball games. In recent months, he has been limping a bit after a vigorous ball game. His owner thought he had a sprained ankle and encouraged him to rest. On a rainy day, they played a ball game in the living room, and here is how it happened. When Max was chasing the ball,

he slipped on a parquet floor and fell over. He howled in pain and hobbled about but still wanted to play, albeit much more quietly, because the pain was overwhelming his desire to race for the ball. After two days, he was clearly still limping, so his owner took him to the vet. He quickly diagnosed a torn ACL and said that the damage to the cruciate ligament had occurred over a long period of time, and that it was now torn irrevocably.

example, to the menisci (one of the two crescent-shaped cartilage pads between the two joints formed by the femur (the thigh bone) and the tibia (the shin bone)) in the knee. This additional injury can result in significant movement restrictions, so the surgeon will be able to examine this during the operation and repair it as well, if necessary.

What is a 'degenerative process?'

During normal movement, a dog's bones, muscles, cartilage and ligaments undergo stress and strain, and – as with every other part of the body – can gradually wear: 'degenerate.' All it takes then is one awkward movement and a worn, damaged ligament could suddenly snap. However, if the ACL has not completely snapped, this is called a cruciate ligament tear.

Cruciate ligament tear is not a question of guilt ("Am I asking too much of my dog?"). My vet reassured me that I shouldn't blame myself; a genetic predisposition may play a role, and some breeds are more likely to tear cruciate ligaments than others. He also told me that within two years of a cruciate ligament tearing in one leg, the other cruciate ligament is more likely to tear also, mostly because the bands of both legs have gradually become 'worn down.' That sounds scary and, of course, we want to do everything we can to avoid that happening!

Osteoarthritis of the knee – what does this mean?

This is premature wearing of the knee: the cartilage is worn away by too much stress and loses its cushioning ability, allowing small surfaces of bone to rub together. In the long term, this will lead to compression of the bone where there is the most weight-bearing during activity. On an X-ray image,

Case history
BASCO

Basco, a high-spirited, playful Border Collie, had surgery on a torn ACL in his right leg six months ago. Everything was going well, and he was recovering his strength. One day, his owner took him to the vet for his annual vaccination. Like most dogs, Basco finds this fairly unpleasant. Although he held still until it was all over, afterward he rushed to the car, and the moment his owner opened the boot, he made a flying leap into the car. This proved too much for his already injured ligaments, and he cried out in pain. The vet rushed to him immediately and diagnosed him on the spot with a torn left ACL.

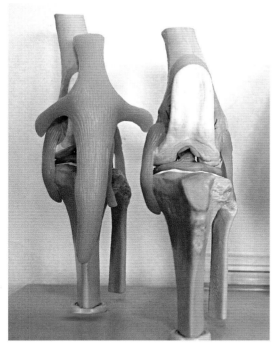

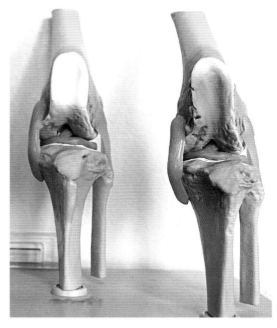

Canine knee joint.
Left: Front view, showing patellar tendon
Right: Front view without patellar tendon and patella.

View of the canine knee from the front.
Left: Mild osteoarthritis
Right: Severe osteoarthritis (cartilage damage with haemorrhage), and clear signs of inflammation, as shown by the red coloration.

the affected areas are slightly lighter than the rest of the bone. Initially, joint mobility is only slightly affected, but since this is a slow, progressive process, if not properly counteracted, eg by a change in gait, muscle-building, etc, the joint can become locked, and, in the worst case scenario, stiffen up completely, leading to inflammation and swelling – and a great deal of pain.

There is no cure for osteoarthritis – all that can be done is to alleviate the symptoms.

Chapter 3

Can I do anything to prevent it?

Yes! There are two main ways you can help prevent a canine cruciate ligament tear:

- weight control
- sensible training

Weight control

I have monitored Argos' weight ever since he was a puppy, because every excess gram of weight is too much – too much stress on his joints and too much of a risk to his health. When I stroke him or comb him regularly, I check that I can still feel his ribs or whether they are hiding under a layer of fat – in fact, he's incredibly greedy and prone to helping himself to food. But, through careful monitoring, he has stayed slim.

From the beginning, I made sure that my dog worked for his daily food ration. He has to really earn every mouthful – there's no such thing as a free lunch in my house! That may sound harsh, but it isn't really. I grew up with dogs in my family, and yet it took some time before I realised that a dog is a working animal. Just like me and probably most people, a dog wants to feel important, to have a job to do, and to be praised and acknowledged for this.

TIP

Argos weighs about 40kg (88lb). We use the scale at the vet to weigh him on a regular basis. If you have a fairly small dog, stand on your bathroom scales with the dog, then weigh yourself without him. Subtract your weight from the combined weight, and you will know how much your dog weighs.

TIP

If you have used a food dummy (a device made of rubber or fabric that can be filled with food/treats) during training, you will know what a good idea it is to let your dog feel useful, and reward him when necessary whilst still controlling his daily food ration. The dummy is a wonderful tool, representating prey which your dog searches for, though, once found, he must bring it to you because he can't open it himself! Only you can unfasten the Velcro to open the underlying zip and, when your dog is (finally) sitting quietly and attentively, reward him from it. It's worth noting that in the eyes of your dog, you are the most important and most powerful person in his life, because you provide the most vital resource: food.

The exercise section includes different ways of using the food dummy, as well as plenty of other exercises you can do, with and without it. However, please do supervise your dog during this activity to ensure he doesn't try and eat the food dummy!

Argos has used the dummy for years; he loves it and is incredibly happy when he sees it – he knows very well that there is food inside and he can hardly bear to let it out of his sight! For more on the food dummy, see *Useful acccessories*.

If the food dummy is not for you because you prefer to feed your dog twice a day from a bowl – no problem. Stick with your normal way of feeding and rewarding, but remember to adjust the amount of food given in the bowl to take into

A food dummy is a great way to reward your dog.

account that given as a reward.

Praise and reward are very important, but, so that his calorie intake doesn't get out of hand, and to ensure the rewards are never boring, I switch between different methods with my dog.

- after saying 'good boy!':
I immediately produce a small toy from my pocket, to play tug with, or ...
- as if by magic, conjure up a ball and throw it a few feet, so he can fetch it and chew on it with enthusiasm, until he has to give it back, or ...
- I get him to sit and wait and then hide a toy that he can search for when I give him the signal, or ...
- I run a bit with him, and when we stop, I give him a quick cuddle (sometimes he looks at me like a fourteen-year-old boy whose mother has just kissed him in front of all his friends!)

However, Argos doesn't get effusive praise or treats after every good performance. Sometimes, I praise him with a friendly 'well done.' It's only when he learns something new that he gets a proper reward. If he has completed his new task perfectly a few times, then I will begin to reward him every other time, then every third or fourth time, and finally only occasionally.

If Argos got a reward at regular

Rewards do not always have to be in the form of food. Toys are a great calorie-free substitute!

intervals, eg, after every third time, he might only do as required every third time because he'd know it was then that he was due something tasty. If, however, he's never sure when – or if – he will get a reward for a good performance, it motivates him to keep working, thinking "Oooh, maybe this time I'll get a treat!"

Does your dog have a good appetite? As a result is he a bit chubby or barrel-shaped? You may think he's adorable just as he is, but try not to feel hurt when the vet says (as he surely will): "Your dog is too heavy" or even "He's too fat." It may hit you hard, perhaps, but he's not trying to hurt your feelings, and at least you know what the position is and can act on it. Your veterinarian can help you determine the ideal weight for your dog and provide a sensible diet plan for him.

Most of the lifestyle diseases of dogs are the result of too much cupboard love, and so are very preventable.

Aisha urgently needs to lose weight so that her joints and

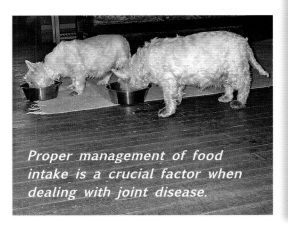

Proper management of food intake is a crucial factor when dealing with joint disease.

ligaments are not unnecessarily burdened, which could eventually result in knee osteoarthritis.

Sensible training

Sensible training and regular exercise will strengthen the muscles, ligaments and tendons of your dog, and build and maintain endurance and performance, which is extremely beneficial, of course.

TIP

Right from the word go, let your healthy dog walk down stairs – this will provide good exercise, and it saves your back if you don't have to carry him. The idea of not allowing your dog to walk up and down steps is now considered an outdated one, and your dog can climb stairs from early in his puppyhood as long as a few rules are observed.

The steps should be the right height so that your dog can go up or come down using one leg at a time without bouncing, or straining himself. Walk your dog on a short lead or harness by your side and make sure that you both use smooth, deliberate steps. (Of course, when choosing the perfect harness, make sure you take into account your dog's size: a large dog will need a different type of harness to a small puppy.)

A properly fitting harness is the best and kindest way to walk your dog. There's a great variety to choose from that are both practical and attractive, like the one in the pictures above. The harness below is a canine 'seatbelt' for use when travelling: a special strap attaches to the D-ring on the back of the harness, and then goes around the car seatbelt when it is clipped in place. The chest strap is padded to absorb shock and minimise the risk of chest injury in an accident. Of course, this harness can also be used when walking with your dog.

Is your dog a typical candidate for a torn ACL?

Caution!

Then he must never race up a set of stairs! He also shouldn't hop – this will hurt his joints, ligaments, and bones. Instead, controlled slow stair climbing (on a lead) is advised – down as well as up (see photos, right).

Watch your dog carefully to ensure he is walking and not hopping, or having to make a huge effort to climb each step. His harness will enable you to keep him under good control.

If your house or flat is spread over several levels, allocate a particular floor for your dog, and prevent him from going down or up the stairs by blocking them. A child's stair gate is ideal for this purpose (see *Useful accessories*).

If your dog is allowed to climb stairs after surgery, make sure you do this in a completely controlled manner as part of his exercise programme.

Incidentally, it wouldn't hurt if you perhaps go out of your way to climb stairs with him; you won't need that gym membership, then!

Incorporate a training plan into your dog's regular walks. This sounds more complicated than it is, as it simply means following these three rules: warm-up – exercise – cool down.

If your dog is raring to go before you've even opened the door of your car, then you will need to make sure he goes slowly during his warm-up. This means spending a few minutes walking more slowly than normal. Once your dog's muscles have warmed up, he can begin doing his training exercises.

Each outing should end with brisk walking on a lead, which will allow the muscles to cool down. You can then give him a massage using gentle pressure, first from neck to tail over the back and sides, with light pressure, then from top to bottom of the legs – he'll love it! (See *Further reading*.)

TIP

The methods used for prevention of canine cruciate ligament tear – ie weight control and sensible training – apply for the period after surgery as well.

Chapter 4

Treatment of the cruciate ligament tear

You are advised to discuss the different treatments with your veterinarian, but, generally speaking, a cruciate ligament rupture in dogs should be treated surgically. You may have heard of a person who was treated for a torn ligament without surgery, but the anatomy of the dog is such that surgery is really the only solution.

If your dog weighs less than 15kg (33lb), meaning that his joints are not subjected to as much stress and pressure as those of heavier dogs, the surgical option should still be carefully considered, as a less invasive technique, such as tightening the joint capsule, may be all that's required.

Immediately after the cruciate ligament is torn, osteoarthritis begins to set in. Studies have shown that if the torn ligament is not operated on, the resultant osteoarthritis will be much worse and will progress more rapidly. But a very small dog's bones may be even more fragile after an operation because they are too delicate to hold the screws in place, for example after TPLO (Tibial Plateau Leveling Osteotomy) surgery. Therefore, you will also need to take into account your dog's bone condition.

Unfortunately, there's no single satisfactory treatment; dog owner and vet must work together to weigh up the benefits and risks of each possible treatment. If a dog cannot be operated on, for whatever reason, the vet will usually bandage the affected leg, and the owner will need to work with their dog on specific exercises to rebuild the muscles, according to his physical capabilities.

Osteoarthritis as a result of a cruciate ligament tear is explained in Chapter 2. The cruciate ligament is now missing so the lower leg has less stability, which results in a greater forward movement, known as the 'shearing force.' This has a negative impact on cartilage and meniscus in the knee during weight-bearing, and leads to other problems, such as arthritis, due to wear and tear.

There are many different surgical procedures for this condition, and your vet can advise you. If he is not an orthopaedic specialist, he will be able to refer you to one. The main surgical procedures used are explained here in this chapter.

If the anterior cruciate ligament (which prevents the tibia from slipping forward out from under the femur) no longer functions properly, then the original positions of the shin and thigh bones will have altered so that the lower leg bone slips away from the thigh bone. On examination, the vet will probably be able to carefully slide the bones back and forth like a drawer, and it is this instability that needs to be rectified.

Firstly, the surgeon opens up the joint capsule (usually arthroscopically, so minimally invasive) to remove cruciate ligament remnants and to identify any possibility of a meniscus injury (a tear in the knee cartilage).

After that, he may use one of the following methods:

Tibial plateau leveling osteotomy (TPLO)

Tibial plateau leveling osteotomy (TPLO) involves making a curved cut in the top of the shinbone (osteotomy), up to the tibial plateau. The tibial plateau (the top of the tibia) is then rotated along the curved osteotomy in order to level the slope (to even it out). A plate and screws are used to hold the tibial plateau in place so that the bone can heal in its new position. During the surgery, the plate is measured radiographically to ensure the best possible fit.

Tibial tuberosity advancement (TTA)

The TTA is a somewhat less invasive surgical procedure which has similar results to the TPLO.

This procedure should not be used for dogs that have a steep tibial plateau (injury to the bottom of the knee joint); therefore, your dog's surgeon will decide which procedure is the best option for your companion. The TTA procedure involves making a cut in the front part of the shinbone, and moving this portion of bone forward in order to realign the patellar ligament (the central part of the tendon), so that the abnormal sliding movement within the knee joint is eliminated. A specialised bone spacer, plate and screws are used to secure the bone in place. Bone graft is collected from the top of the shinbone and placed in the gap in the bone to stimulate healing.

Capsular plication and ligament replacement

The aim of this procedure is to bring the capsule of the joint (the connective tissue, various ligaments and muscles which surround the joint and hold it together) closer to the bones of the joint, making the joint tighter and reducing joint volume and looseness (laxity). It has the additional effect of improving joint proprioperception (feedback to the muscles from the nerve endings in the joint). In short, the capsule doesn't allow the bones any free play, or room to move away from one another. This method can also be supported by an extra ligament outside the capsule.

Every joint has a capsule around it which consists of fibrous material and is filled with joint fluid (synovia). In the method described, this capsule is made smaller and tightened by the suture and following cicatrisation (scar tissue formation), so the joint inside has less opportunity to move outside of its normal range and thus becomes more stable (and less prone to cruciate ligament injury).

Of course, your vet will do everything possible to prevent your dog from developing severe osteoarthritis. As a reminder: osteoarthritis is a degenerative disease. Once the ligament has been torn in the knee, osteoarthritis will occur because the body is reacting to the unjury. If the meniscus are also affected, this may lead to problems during the healing process.

Visit Hubble and Hattie on the web: www.hubbleandhattie.com and www.hubbleandhattie.blogspot.com
Details of all books • New book news • Special offers

19

Chapter 5

Before the operation

Ask your vet about dietary supplements such as products containing green-lipped mussel extract, which helps to stabilise tendons, ligaments, and joints, and slow the onset of osteoarthritis, or prevent it from worsening too quickly.

Medication

The vet will prescribe painkillers which you should give to your dog according to the instructions so that he won't be uncomfortable or in pain. Being comfortable is an important part of the recovery process, because long-term pain puts a lot of stress on the body, and could cause your dog to move in a way that is unnatural, in order to avoid pain, which could result in further damage.

So, ensure that you keep your dog almost completely free of pain ... Why only 'almost?' Because complete pain relief would be dangerous as then your dog would have no warning if he was moving in such a way as to cause further harm to himself.

Of course, it's your responsiblity to ensure that you do not allow your dog to do anything that could cause more harm, such as jumping on or off the sofa – or chasing the postman!

It's a good idea to ask your veterinarian about homeopathic remedies, which are available as:

- drops
- tablets
- globules (which are like beads), or
- small globules (very small beads which have less space between them, make more mass and soak up more liquid)

Caution!

The active ingredients of homeopathic remedies enter the blood via the oral mucosa (the mucous membrane of the mouth, including the gums). If they are mixed with food, they will not remain long enough in the mouth to be absorbed, and so will not be effective.

Homeopathic tablets and pessaries have a coating made from milk sugar (lactose) which contains the active ingredients.

Caution!

Some dogs are lactose intolerant, and even very small amounts can cause diarrhoea. If you're unsure about how your dog may react, initially try a very small dose to determine whether he can tolerate the tablets or globules. If he gets indigestion, talk to your vet.

Help! How do I give my dog medication?

Administering homeopathic medicine shouldn't be too much of a problem, simply because it doesn't smell or taste unpleasant.

You can crush tablets and pessaries between two teaspoons and, using moistened fingers, apply this to the oral mucosa (inside of the lips, under the tongue, along the gum). If your dog finds this approach a bit strange, you could practice a few times with a tiny dab of liver paste, for example, and a few little treats ("yum, this is nice!") until he gets the idea that your finger in his mouth equates to a nice, tasty treat.

Another option is to smooth some yoghurt over the surface of a bowl, sprinkle the crushed tablets on top, and then allow him to lick the bowl. In this way, the mucous membranes can absorb the active ingredients before they make their way down the oesophagus.

You don't need to crush the globules because they're small enough, and can be administered in the same way as the crushed tablets.

Drops can be given either neat or diluted with a little water using a disposable syringe (without needle) directly into the mouth. If you are unsure about this, practice it a couple of times with a small amount of soup or similar. If the sight of the syringe causes your dog to wag his tail in anticipation, you know you've won!

As for conventional painkillers in tablet form, depending on your dog, these are often not as easy to administer.

Aisha will eat anything, including pieces of bread spread thinly with butter which have a tablet hidden within. Basco will gulp down pills if they are wrapped in a slice of sausage. Argos is so keen on cat food that he doesn't even notice there's a tablet in it! Donna, on the other hand, is not so easily outwitted, and has formulated a clever technique. She will move her mouthful of food around on her tongue until she has isolated the foreign object and then spit it out.

Sometimes it helps to give your dog four or five identical small pieces of food. Conceal the tablet in the third piece of food. Your dog will eat the first two pieces, and, reassurred that there was nothing untoward hiding in them, will take the third offering with little suspicion, rushing to eat it because you have a delicious fourth snack ready, held high above his head so he has to stretch up to reach it and gobble it down ... easy!

Any tricks are allowed – as long as you keep it friendly!

If nothing seems to work, you may feel frustrated – but talk to your suspicious or hesitant dog in a friendly tone. Reassure him that he is the greatest dog in the world, your absolute favourite, a really lovely, wonderful dog! Here's what to do next.

Have your dog sit next to you, preferably where there's little chance of escape, such as in the kitchen next to the table, on which the tablet is hidden in a small piece of sausage, wrapped in a large portion of cat food, or clamped in a tiny piece of cheese.

Talking reassuringly to your dog the whole time, hold the top of his muzzle and gently lift his upper lips until you can see the space between the upper and lower teeth (A).

Extend your fingers into this gap and gently open his mouth just enough so that you can push the disguised tablet as far back in his throat as possible (B & C).

Hold his head pointing upward (but not too far back as this will scare him), and stroke him gently from the jaw down to the throat to promote swallowing (D).

Praise your dog when he swallows – tell him what a good dog he is, and give him an extra treat as a reward.

Caution!
Give your dog only very small portions of sausage and cheese. Sausage contains a lot of fat and calories, so, the smaller your dog, the smaller/lower in calories the 'packaging' of his tablets should be.

Contacting a canine physiotherapist before surgery
Seek out a canine physiotherapist in your area and have a chat with him or her. Generally speaking, you will receive a recommendation for physiotherapy treatment from your vet or the surgeon after your pet has been discharged.

So, now you're bound to be

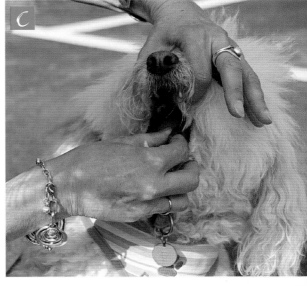

wondering, why is physiotherapy so important? Any interference with a joint puts stress on all involved tissue structures. During the operation, the surgeon will have moved or stretched muscles, tendons and ligaments (and also removed the remains of the damaged cruciate ligament); the joint capsule will have been opened, and so on. Following surgery, structures and tissue in the area will be traumatised: a violation such as this sets off a chain reaction in the body, which needs to be treated and then gently worked to restore performance. This speeds up the healing process and minimises any damage.

Introduce your dog to the physiotherapist slowly, without causing him any stress or anxiety about the new environment. He should get to know the therapist ("nice lady") and become familiar with the practice ("smells like dog-interesting stuff"). When, later, he has his operation, after all the strange new things that have happened, he will have something familiar that will help to rebuild his trust in the world again.

During the familiarisation process the therapist may lead him to the trampoline or the fabric tunnel, for example, give him a chance to sniff everything, and maybe have a go on one or two things. He might enjoy it so much that after his operation and after the stitches have been removed, he looks forward to going back to this fun, (dog-)friendly place.

It is also reassuring for you to know that your dog will be in good hands.

The physiotherapist will tell you what to expect in terms of therapy costs. If you budget your time and money, you'll be able to plan your next steps for the weeks after the operation.

For the period before the

(and opposite, top) The tunnel is used as a training aid.

operation, you can use additional measures such as cold/hot packs, (see *Useful accessories*) and give him massages. If you're unsure how to do this, buy a good book that will explain and demonstrate the techniques, and advise on how they should be used (see *Further reading*). You'll also feel better if you can actively do something useful for your four-legged friend.

Since you must now limit the length of the walks you take, it's important that he is exercised passively. Passive motion stimulates the metabolism in all tissues involved in movement, and promotes blood circulation. Improved blood circulation supplies the cells with more nutrients, which, in turn, helps prevent stiffening of the joint capsule. Again, a good book such as *The Complete Dog Massage*

Manual, published by Hubble and Hattie, has all the relevant information about this technique.

To be on the safe side, ask the physiotherapist to show you how to support your dog without causing any harm during passive movement massage.

On the lead

Your vet will have advised that you keep your dog on a lead whilst out (and just a one metre lead at that!), and walk only short distances, such as to and from the local shops. Stay away from other dogs to avoid joyful reunions. Let doggie friends know about your dog's impending operation so that they understand why you can only greet them from a distance.

It is very important to prevent your dog from jumping into the car. Jumping out of the car or from somewhere high (chair, low walls, table, etc) is also off limits, as it will place stress on the front legs, shoulder, and all other joints. And remember: because your dog will favour his hind legs, the front legs will have to do more than normal anyway.

Now may be a good time to get a car ramp (see *Useful accessories*), as this will not only be a big help with getting your dog into and out of the car, it also means that grooming can be done on a table instead of on the floor. At the same time, with the non-slip board, you've also got a training aid for uphill running.

Maybe your dog is a small Yorkie whose favourite spot is on the sofa ... from now on, jumping from places such as the sofa, bed, or chair is prohibited. Lift your little dog up or down, or get him used to lying on a cosy dog mattress on the floor, or provide him with a small ramp (see *Useful accessories*).

Nutrition

Due to the necessary exercise restrictions, your dog's calorie requirements will decrease, and he'll need less food. Your vet can advise on the best way to reduce food intake, without causing any nutrient deficiency in your dog.

Remember to ask for products containing green-lipped mussel extract – this helps to slow the onset or worsening of osteoarthritis, and is beneficial for tendons, ligaments and joints.

Fill a large Kong® toy with food (for more on this, see *Useful accessories*). He will have to use his brain to figure out how to get the food, and this will occupy him

Kong®: mentally and physically entertaining for your dog!

for some time. Again, he is having to work for his reward. Use cottage cheese with apple, pear or grated carrots instead of calorie-laden foods such as peanut butter.

Aisha's owner has a food processor in which, every morning, she processes apples, pears, cabbage, and/or carrots. The mixture is kept in a jar in the fridge, so she always has a low-calorie treat available. Sometimes she fills the Kong® with apple and carrot

TIP

Can you replace part of his food allowance with apple slices, pear or carrots? Perhaps you can make extra-thick, extra-hard baked wholemeal biscuits, using copious amounts of apple or carrot in the mixture?

gratings and a spoonful of low fat cottage cheese, so that she doesn't feel as if she is depriving her dog!

Tomorrow is the day of the operation

The vet has advised that your dog should fast before you take him to the clinic. As a precaution, check when his last meal should be and when he should stop drinking water, prior to the operation. Be sure to follow your vet's instructions to the letter!

But why must your poor dog starve? Surely he has been punished enough already?

Of course, he is not being punished, only temporarily going without food. The fact that most dogs will do anything to get food is due to their basic survival instincts (eat now, while there's something to eat as who knows when the next meal will be?). If you follow the guidelines from your vet, nothing terrible will happen to him – on the contrary, this will keep him from harm.

Why is this strict fasting necessary?

Your dog will be given a general anaesthetic, which may cause him to feel, or actually be, sick. If he has food in his stomach whilst unconscious, he may suddenly vomit. He won't be able to get up, like he can when he's sick on the living room carpet, and adapt his breathing. Therefore, there's a very real risk that he may choke on his own vomit. Should this happen, the surgeon might be able to abort the operation in time to avoid the worst – but maybe not. In any case, such a scenario puts a considerable strain on the dog. And, of course, the practicalities are that the surgeon will have to cancel the operation, book another appointment, administer another anaesthetic, etc, etc … and your vet probably won't be very happy, either …

So, fasting really does mean fasting!

You will feel terrible as your dog stares sadly into his empty bowl, but stay strong, and make sure that the rest of your family supports you with this. For your dog's sake, please don't be tempted to give him 'just one treat'; it could well be his last if you do.

Chapter 6
After the operation

So, the operation has gone well, and now you can collect your dog. If you've not already done so, you will need to sort out how you will transport him home. Ask the surgery if he is still very drowsy, and what sort of behaviour you should expect (extreme fatigue, apathy, loss of appetite, etc), and over what period of time. Clarify how often the dressing should be changed, and make an appointment to do this.

All dogs are different, of course, and may react in various ways. For example, Aisha had no appetite and wanted to be left alone. All she could manage was a few steps to the nearest patch of grass to do her business.

Argos was quite different. He wanted to be near me, languishing in self-pity, and looking as if he were half-starved and generally neglected. He kept leaning against my leg, his posture emanating very deep sorrow.

Back to normal
You may find that small family members, in the mistaken belief that it will help, will lie down next to your dog, stroking and touching him, feeling sorry for the poor, dear creature, and giving him bits of food.

Pampering like this is counter-productive, however. You may even find that your dog has previously hidden thespian talents, and really gets into the role of tragic hero, bribing you to feed him lots of treats, which will eventually ruin his health.

Of course, your dog has been through the ordeal of an operation – but he doesn't need pity, only for his family to treat him as they usually do, with lots of love, care and respect. You don't need to compensate for his (temporary) lack of joie de vivre. Similarly, you don't have to reward him with extra treats because he has had an operation.

The less fuss you make, the more likely it is that your dog will feel that things are normal. All you need do is whatever is necessary to help him heal: make sure he's not in pain, is not hungry or thirsty, or suffering from boredom.

The vet will give you advice about how to help your dog on his journey to recovery. He will tell you what drugs have been given to him during and after the operation, how long the painkillers take to work, when you should administer them, how many, and at what time intervals. Be sure to follow his instructions precisely.

More medication

"Does Aisha really have to take painkillers? She's bound to get used to them after a while and then they will have no effect."

"What if he ends up addicted to the painkillers?"

You are allowed to feel sorry for your dog after the operation, but don't overwhelm him with treats and attention.

"I am apprehensive about the many side effects of the tablets, all of which are listed on the packet! How can I do this to my little dog?"

"Pain is a necessary part of life, isn't it?"

The ability to feel pain is essential for survival. Picture the following situation: a bubbling saucepan full of chicken soup, giving off tempting smells which your dog just can't resist. He starts drooling with anticipation, gets his front paws up on the stove, and begins lapping up the soup, which he gulps down. The resultant burns and internal injuries would be pretty horrific.

Pain can prevent these and other injuries from occurring: pain protects him because it warns him in advance that something is wrong. In the case of the soup on the stove, the warning is: "ouch – hot – stay away!"

However, long-term pain can cause a huge amount of stress, over and above the everyday, the-children-are-driving-me-mad and I-never-have-time-to-do-anything kind of stress. Persistent pain causes a kind of stress that negatively affects the entire body and, for example, upsets and disrupts metabolism. It can suppress the immune system, reduce respiratory function, and interfere with wound healing – to name just a few side effects.

Aisha, Argos, Basco, and Chico should experience as little pain as possible, not only because pain is unpleasant, and because no one wants their beloved pet to suffer needlessly, but also because chronic pain in animals has hugely negative biological consequences.

Another important factor: if Aisha's leg is causing her pain, she won't want to move about. Without painkillers, she would only get up for the most essential things, and then retreat quickly back onto her blanket. Argos would do the same because he doesn't want to suffer any unnecessary pain. When dogs are injured, they are usually very quiet, unassuming patients, and easy to care for, though unwilling to exercise if in pain – not for all the love in the world! But it's exercise which will help them heal over the coming weeks.

Muscles can be built – or rebuilt – only by repeated resistance training; that means regular training of the muscles or muscle groups, using guided, purposeful movement.

If you find a movement or action very painful, you probably won't want to do it – so why should your dog? However, you are responsible for helping your dog do the exercises that the physiotherapist and the vet will recommend to you, as well as ensuring that he is not in pain. Give him the painkillers prescribed by the vet – they are

TIP

TIP
Give your dog his morning dose of painkiller well before the first walk, so that it has plenty of time to take effect before you go out. The vet can tell you how long this is likely to be, and for how long each dose will be effective, which will allow you to determine the best time to take your dog for a gentle walk.

essential to the success of your efforts and rehabilitation measures. And as he gradually recovers, progressively discontinue his painkillers.

Bandages and dressings
If the vet is going to change the bandage or dressing, then you need only ensure that the dressing remains dry, which represents quite a challenge in wet weather!

Ideally, you will need a protective shoe that is sufficiently large and offers reliable protection in bad weather – see *Useful accessories* for more information.

Possible complications
Argos had to wear a bandage for quite a long time after his operation. Because the canine leg has a bigger circumference at the top (in common with most species), the dressing sometimes slid down a little and then came loose. This caused the dressing material to become 'rucked up' in places, and then press on the fur and the skin, resulting in some very bad wounds that had to be treated at each of the frequent dressing changes.

Dressings shouldn't cause wounds, of course, but there's always the risk that a wound could become infected, so it's vital that either you or your vet change the dressings regularly, and keep an eye on how well your dog is healing. Talk to your vet about treatment options, and ask the physiotherapist what you can do to help your dog heal more quickly. He or she may recommend the use of low-level laser therapy or therapeutic ultrasound, which will significantly

TIP
You can improvise by wrapping your dog's paw in a plastic bag and securing it with parcel tape. Then put a waterproof surf shoe (a special waterproof 'shoe' made of neoprene) over the top and secure with tape to prevent the whole thing from coming undone.

speed up wound healing (both methods are described in more detail in the chapter *What can the physiotherapist do to help?*).

Your vet will be happy to explain to you the various operation procedures, and which is the best for your dog. Even after a successful operation, however, occasionally complications can occur.

The joint may become inflamed for various reasons, ranging from over-burdening the joint to bacterial infection.

Signs of inflammation may include:
• your dog favours his bad leg, and may only use the other three when walking
• he has a fever
• the joint feels warm, or even hot
• the joint is swollen
• when you gently move the joint, you can hear rubbing noises and, in more advanced cases, a cracking sound as air escapes from the tissue in the joint (the vet may refer to it as crepitus)
• there is pus around the inflamed area

In a TPLO (Tibial Plateau Leveling Osteotomy), a special surgical procedure, a small metal plate (implant) is screwed to the shinbone. Another type of operation, which also involves implants being anchored to the bone is called a TTA (Tibial Tuberosity Advancement).

After the operation, you may find that, at very low temperatures, your dog appears to suffer a sudden sharp pain in his leg, or howls in pain when swimming, or when you put a cold pack on him to treat post-operative swelling. Your dog's sensitivity to cold will increase after an operation such as this as metal is a good conductor of temperature. Winter, generally, won't be an issue for him, but be careful in very cold weather. Try to determine what he can and can't tolerate.

In some cases after a TPLO/ TTA, the animal cannot tolerate this foreign body, and begins to show signs of rejecting the metal plate.

How can you tell if this is happening? The evidence of this include the signs of inflammation described previously, and could mean another operation to remove the plate. Your vet will then have to decide what steps to take to help your dog.

Visiting the physiotherapist

This is what a physiotherapist will probably do and ask on the first visit:
• give your dog time to get to know them (to sniff) them
They will also go through the history of your dog's injury:
• what happened and how?
• which procedure was used to operate on the dog and when?
• who did the operation?

- who your vet is and what is his contact information for any medical issues?
- are there any other health problems, allergies, intolerances, etc?

The physiotherapist will analyse your dog's gait: ie how he walks, whether he is trying to avoid walking on his injured leg, or if he hobbles or limps. She will measure the leg circumference in order to assess muscle strength in order to evaluate any progress or setbacks. She will test muscle reaction, and assess his balance.

She will create a treatment and training plan by consulting with you, and she will show you how to keep a training diary.

This diary doesn't require much effort, but is a good resource for you to record your dog's progress, from week to week, or even day to day. It makes it easy to identify any issues that need particular attention, and what kind of exercises you should focus on with your dog.

Care and activities during the rest period

Your dog probably won't enjoy being inactive. If Basco has nothing to keep him amused in his waking hours, he soon becomes bored. A dog who doesn't have enough to keep him busy may feel unhappy, depressed, irritated, grumpy, or rebellious. Or he may try to amuse

---

Case history
ARGOS

I kept a daily record of which exercises worked well (very good, average, poor, easy or difficult, etc), which ones Argos got better at and those he didn't. For example, after surgery, it took him a while to balance on a tree trunk again, which before surgery he could do very well and for quite a length of time. After surgery, however, he could take only a few wobbly steps. When he was finally able to move forward two or three feet on the tree trunk, I felt incredibly proud of him!

I referred to my notes when we went to physiotherapy: that way, I could advise the therapist which exercises Argos was very reluctant to do, or couldn't do at all, which provided important clues to how he was healing. The diary showed progress which I may not have noticed if I hadn't written it down every day. I am convinced that Argos could sense my triumph and happiness, which was a great motivator for him as well.

himself – which is rarely to the liking of his humans!

Goldie, a five-year-old Labrador bitch, turned her attention to the legs of some antique furniture to relieve her boredom. Gecko, a six-year-old sheep dog cross, ripped off part of the wallpaper and carefully removed the carpet strips on the living room floor. Dogs need something to keep them occupied and, immediately after the operation, a variety of exercises and mental challenges. It is important that you give your dog something to do. Aisha, for example, finds foraging for food highly satisfactory.

Here are some game ideas:
- let your dog search for toys which have been hidden in the house, put food scraps in a sealed toilet roll, or hide them behind half-open doors
- hide a treat under a saucepan

Many household objects are great to use in 'find' games.

lid and see if he can lift the lid to get it
• put some scraps of food in an old sock, seal it with a rubber band and hide it – your dog must find the sock and bring it you before he (every second or third time) gets a treat out of it. Have him sit before sending him to search for it, then sit again when he gives you the sock – the action of sitting/standing, standing/sitting will strengthen his muscles

Could your dog learn something new? Perhaps a trick, which you can perfect indoors? You could rehearse for a show at the next children's party, or Grandma's 80th birthday; for example:
• teach your dog to pull off your socks, gloves, or cardigan (use old items to practice with)
• if he can bark on command, practice the trick 'my dog can count'
• teach him which container is the bone, ball, or treat hidden in?
• sit on the floor and let your dog search for a toy concealed in your sleeves, trouser leg, pocket, etc)
• teach him the 'touch' trick, where he has to touch different objects with his nose: your hand, the ball, the door … You could even get him to operate a light switch
• he could learn to pick up and carry things in a basket, such as

the newspaper or letters from the doormat
• he could put socks and other small items of clothing in the washing machine
• how about if every toy has a name so you can ask him to bring Kong®, bring Teddy, bring the ball, the duck, the frisbee, etc
Rico was trained by his mistress to bring 77 different objects. After he had a shoulder operation at the age of nine months, he needed to rest. In order to challenge him mentally, his owner began to teach him various words. Within nine years, he could recognise about two hundred words.

Of course, not every dog will be able to do this, but Rico's example shows you how you can put your dog's mental capacity to good use and keep him occupied at the same time. Dogs need stimulation so that they don't become depressed or bored.
You can also buy or make intelligence games for dogs; see *Useful accessories*, where you will find information for some game ideas.
You could also fill the Kong® with tasty, low-calorie treats, which he will spend some time getting out.
Get your dog moving – this is important for his heart and circulation, muscles – and his soul. The vet will have prescribed strict

my DOG has a cruciate ligament injury . . . .

During his recovery time, you could teach your dog all sorts of useful tricks, such as how to put his toys away, as well as get them out!

rules regarding on-lead exercise only, so your dog is still not allowed to run or jump, or play with his canine friends for the time being.

Just how much your dog can do depends on the surgical procedure undertaken and the healing process. Talk to your vet and/or physiotherapist about this, ensuring that you follow their instructions.

This 'treat ball' is just the thing to provide mental and a little controlled physical stimulation for your dog. He'll love pushing it around with his nose, trying to get the treats to drop out!

Chapter 7
What can the physiotherapist do to help?

The physiotherapist offers a range of techniques to treat your dog according to his needs, to promote healing, build muscle, strengthen ligaments, and relieve tension. He or she will also explain and demonstrate which of these techniques you can use at home, and how to do them safely. And you may even be able to borrow some equipment to do the exercises at home.

The most commonly used techniques used are:

Massage
Massage is very relaxing, relieves tension, and thereby eliminates muscle pain. Toning techniques encourage blood circulation and increase oxygen supply to the muscles. One massage technique, effleurage, is particularly good at helping to re-establish mobility.

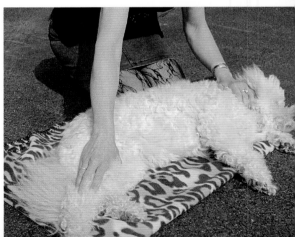

Heat therapy

Heat improves circulation, increases metabolism, and relaxes the muscles. It will also make the connective tissue (eg joint capsules) more flexible, which increases mobility. The physiotherapist uses infrared lamps or hot packs (see *Useful accessories*) before exercise therapy.

Cold therapy

Reduces swelling and inflammation (for cold packs, see *Useful accessories*).

Ultrasound therapy

While hot packs and infrared radiation heat the outer layer of the body, ultrasound reaches the deeper layers. The heat generated there reduces muscle tension and pain, and improves circulation, which in turn speeds up the healing process. This therapy improves joint function, and some of the existing stiffness will ease. Depending on the sound intensity used, penetration depth is 1-5cm, which will have a beneficial effect on the joint capsule.

Top left: Your hands should form a soft and uniform shape during effleurage.

Left: The positive influence that massage has on the circulatory system can help promote healing.

Electro-therapy (TENS)

TENS (Transcutaneous Electrical Nerve Stimulation) stimulates the skin using mild electrical impulses on the nerves, and blocks pain signals before they can be received by the brain. You may have seen this device already in your physiotherapist's practice, as it is also used on humans. During a treatment, you will see that the muscle in your dog's leg gently twitches, possibly relieving tension and easing pain, allowing him to relax enough to doze for a while. These minimal movements help build new muscle tissue, and thus prevent muscular atrophy (muscle wasting). At the same time, the process will regenerate nerve tissue and bone.

Due to a small change in posture and unnatural weight-bearing, Aisha has tension in her back. You probably know how this feels if you have ever suffered from backache: it hurts.

Dogs howl if they feel a sudden pain. With longer-lasting pain, however, they suffer in silence, and, unlike us, can't say where it hurts. If the physiotherapist can reduce or completely eliminate muscle tightness by applying brief electrical stimulation, this is such a relief for Aisha.

It's possible you may be able to borrow this piece of equipment to treat your dog at home.

Additional applications for this

TENS therapy stimulates the nerves using mild electrical impulses.

type of stimulation range from muscle strengthening to therapy for paralysis. These treatments should, however, only be carried out by a trained physiotherapist.

Low-level laser therapy – intensive light therapy

The high energy laser light used in this treatment penetrates the subcutaneous layers of skin, and works deep down in the muscles, which provides pain relief and improves circulation. It also stimulates the metabolism, which means that by-products of inflammation, such as toxins, are removed from the cells more efficiently.

If your dog has an oedema (an accumulation of fluid in the tissue), laser therapy reduces swelling, because it aids lymphatic circulation and lymphatic drainage. In addition, the laser light is directed at certain acupuncture points, which aids the healing process. Laser therapy is also used to heal scar tissue.

Exercise therapy (active and passive)

For active exercise therapy, the therapist designs exercises for the dog which require movement of specific, individual muscles and muscle groups. In passive motion therapy, the dog does not move of his own volition; the therapist moves parts of his body – such as paws or legs – for him.

The ultimate aim of exercise

therapy is pain-free mobility of the dog, or, at the very least, to maintain and build muscle, give him the desire to move (again), and create and improve physical fitness and endurance.

Aquatherapy
If you've ever done water aerobics you'll know why it's so good for you – the buoyancy of the water means your joints are protected, and you almost feel weightless. A dog experiences the same thing during aquatherapy.

The pressure of the water also acts as gentle lymphatic drainage, and therefore reduces swelling. Water resistance promotes muscle growth, and if the physiotherapist

Laser therapy is relaxing, reduces pain, and improves circulation.

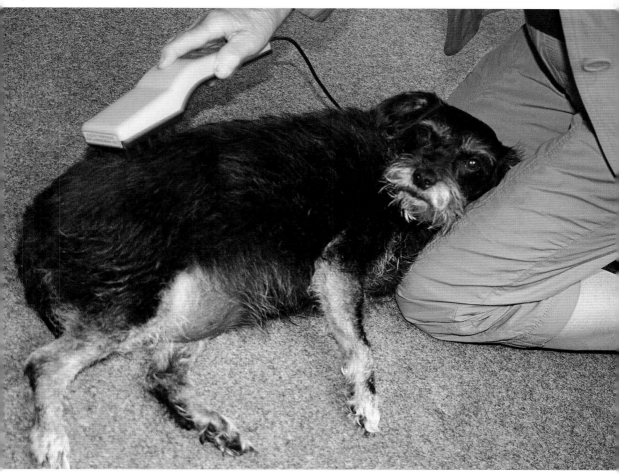

turns on the jet stream for a few minutes, your dog will love it!

Water pressure provides resistance, improves circulation, and boosts the metabolism, which will all help your dog to heal quicker.

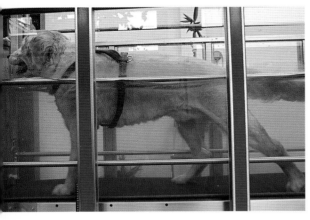

Exercising on an underwater treadmill strengthens your dog's joints and muscles ...

... and improves his posture and gait.

Scar treatment is a major concern for the physiotherapist. Why? Well, to a dog, scars are probably of no great importance, but scar tissue is less elastic than normal skin, and it may be lumpy and hard. This can cause problems such as:

• itching and inflammation
If a scar itches, your dog may try to gnaw on it, and it may become infected as a result. Your dog may need to wear an 'Elizabethan collar' after his operation

• pain
Scar tissue is less flexible than the underlying layers of skin. With movement, the flexibility of the different layers vary – normal skin is elastic, while scar tissue is more rigid, which can cause your dog a lot of pain

• licking
When dogs lick their skin, they can cause eczema. Many people believe that dog saliva is healthy, when, in fact, it's quite the opposite! And because the skin stays damp from licking, it's an ideal breeding ground for all sorts of pathogens

• raised scars
The thicker the scars, the more of a problem they become, because of their inflexibility and low elasticity,

and the increased risk of pain, leakage, etc

- allergic reaction to suture materials

Just like people, dogs can be allergic to a substance – even to the stitches which hold the wound together. This will obviously hinder the natural healing process. The physiotherapist will help to identify such an allergic reaction early on, so you can be advised about what to do next

- encapsulation of residual suture material

If the 'seam' of the wound is not fully knitted together, the body's immune system will try to encapsulate the stitches in order to break them down. Ultrasound therapy and/or an intense scar massage will help the body reabsorb the suture faster

The physiotherapist will massage the scar with the aim of achieving optimal flexibility in the skin tissue, and giving pain relief. She will show you how to do this so you can help the wound heal successfully.

Chapter 8

After the stitches have been removed

Does my dog still need his medication?

Yes, definitely. Building muscle and exercise training are only possible if your dog is mostly pain-free. Discuss this with your vet, and also ask about possible alternatives, such as homeopathic remedies.

As previously mentioned, food supplements containing green-lipped mussel extract help to prevent the development of osteoarthritis. Make sure that your dog's food contains enough high quality protein, which builds muscles. Each dog needs a different amount, depending on his age and general health, so ask your vet for his or her recommendations.

Caution!

Dogs cannot digest raw egg whites – you're better off using these to bake coconut macaroons, but only for your human family members!

Off the lead – what now?

At last, the vet has decided that the wound has healed. But that doesn't mean your dog can race out of the house, jumping for joy! All of the new-old movements must be tackled in a calm and measured manner.

• ask your vet how long you can walk with your dog, at what point you may extend this time, and by how many minutes

• watch your dog as he walks or trots beside you, and keep an eye on whether he limps to protect one leg (which one is this?)

Gradually, you'll be able to walk for longer, and eventually you can go jogging with your dog. You might even be able to cycle slowly on your bike, while your dog accompanies you alongside. However, he shouldn't be allowed to gallop.

It's tempting to involve him in a

game of frisbee at your local park, and let him enthusiastically greet all his old doggie mates, but please supervise him closely at all times, to make sure he doesn't get hurt.

Obedience problems after a long time on a short lead

Sometimes the period of rest can seem too long for a dog, especially if he has always had a strong urge to be active, and a rich and eventful social life. You may be amazed to find that bad manners have suddenly appeared from nowhere: for example, Goldie pretends not to hear the usual whistle, and has developed a sudden penchant for hunting.

The following is an example of an exercise I use from time to time if Argos is in very high spirits, to help me get him under control:

I put Argos on the 5-metre long lead. Not looking at him I take 20-30 steps in one direction, stand still, and still do not look at him (which would otherwise be an invitation to play, and he must earn this first). I wait until he comes to me, sits down and makes eye contact – and at that exact moment I say "good" and we play tug of war with a toy, or I toss him a little treat. Once he has finished the treat, I take (again without eye contact) 20 to 30 steps in a different direction, stand still, wait until he sits and watches, then I

praise him, and play a quick game or give him a little treat – and so on, again and again.

If I'm standing still, and Argos won't come to me, I wait 30 seconds and walk 20-30 steps again. Sooner or later, he will come over to investigate! Pretty soon, he realises that if he doesn't sit still and look at me, then nothing nice will happen!

If he barks, this is his attempt

With your dog on an extending lead, walk 20-30 paces away, and stop. Without looking at your dog, stand and wait ...

... until she comes to you, sits, and makes eye contact with you.

to persuade me to play or to give him something to do, or whatever. I stay silent and just ignore him, look in another direction or, even better, turn my back on him and move on until he is quiet. If he is quiet and sits and looks at me, then we play a quick game or I give him a treat.

After a while, he begins to understand the ritual and then we can play for a little longer. I let him play with a ball (but no further than the length of the lead), or we run for a little distance. Sometimes, there is even a word of praise, which increases his motivation and willingness to focus on me ("When will she finally give in?"). He doesn't want to miss a play session or a morsel of food, but he never knows when or if it will happen.

As soon as he shows signs of wanting to be off the lead, I immediately put him back on a short lead and walk briskly with him, with sudden turns, and tight corners, asking him to 'heel' and 'sit' along the way. Once he has calmed down, I put him back on the long lead.

Granted, I sometimes feel like a Sergeant Major, but the only way to ensure Argos re-learns his obedience training is by being consistent and persistent (you are allowed to sigh here!).

Case history
ARGOS

Argos loves egg yolk more than anything else, and has one three or four times a week – but he has to work for it! Before his operation, he could sit with his front paws raised to get his treat. After surgery, he learnt to do it again, but with difficulty. Gradually, he has managed to build up his strength and can hold his balance in this position for up to ten seconds.

Above all, I make sure I keep my dog incredibly busy with a variety of exercises such as dog-dancing, search games and scent-tracking games (see *Useful accessories*) – these are all activities that he loves and which demand his full attention. This shows him that it is much more fun spending time with me than trying to run off, and he gets plenty of exercise and his mind is kept active. If he runs off, he can never be sure that I won't simply ignore his running away – there is sudden panic when he turns around and I've suddenly disappeared with the food! (the food is probably his main concern!)

Of course, it may be difficult to retrain a dog that has just been 'released,' and you may feel sorry for him, thinking: "He just wants to have his freedom ... enjoy life ... run around properly like he used to." But if you let these bad manners become a part of everyday life, it will be very difficult afterwards to regain your authority and convert the bad habits, however understandable they may be, to acceptable behaviour.

Quite apart from that, chasing buses is a very dangerous activity ...

Get your dog's attention – and calm her at the same time – by walking briskly on a short lead, making sudden turns, and asking her to 'sit' and 'heel.'

Chapter 9
Aquatherapy

At last, the wounds have healed, the stitches have been removed, and the vet has no objection to a spot of aquatherapy. Perhaps you have heard of the underwater treadmill but aren't quite sure what it is?

After a detailed conversation with the therapist, and possibly an opportunity to explore the actual treadmill, you'll see that it really is quite an impressive device.

Like most Retrievers, Aisha is excited about anything that is wet and has to do with water, and finds the warm and humid smells tempting. But what a large tank full of water, technical stuff everywhere, and strange noises! Aisha is extremely curious!

Dr Häusler advises at her practice:

"For the first treatment, it is important to reassure the animal

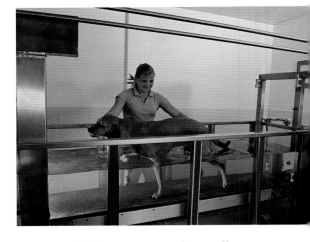

Underwater exercises offer many advantages, including greatly improved mobility.

and owner alike, because they are both bound to feel anxious and unsure about the new situation. For the first visit, I recommend taking something along for the four-legged patient – this could be a treat or a dearly-loved toy. You could bring a few different things and conjure them out of the bag, so that your dog considers this occasion exciting and not frightening. The first treatment will vary from dog to dog. Duffy has no reluctance about getting on the treadmill. Dart is completely different – he needs time and lots of encouragement. Dart's mistress finds it frustrating that her dog refuses to climb the steps to the treadmill. 'He's not usually like this,' she complains impatiently.

Rule number one: Keep calm! "It doesn't matter," explains Dr Häusler to her, "Dart is a careful dog, which is a really positive thing. An impulsive dog could be in danger of injuring himself. Over time, and with encouragement, his anxiety will lessen; if he comes in all stressed and tense then this is counterproductive. A tense body cannot train properly, and could even suffer injury. He would also return home with the impression that this is a nasty place, where he experienced bad things. And next time, he would be in a state of anxiety before he had even got in the door.

"At last, we manage to get Dart on the treadmill. I let it go up first and then move it sideways across the pool. Finally, I let it down very slowly, until Dart has his feet in the water. Friendly persuasion and lots of praise reassure him that everything is okay. Toys and treats are an excellent distraction. Next, I sink the treadmill bit by bit into the water. It is fairly warm, so Dart isn't cold – yet it takes him a while to realise that all these strange things aren't a threat to him.

"The deeper the dog is lowered into the water, the greater the support for his joints. This is all new for him, and he can't control the motion of the treadmill. Gradually, however, he will begin to feel relief from pain. When the treadmill begins to move, he is puzzled at first, and then very soon realises that he can actually run on the strange surface – and is praised effusively for it.

"This first session with the underwater treadmill is not so much about training, but merely to reassure the dog that he will come to no harm. These early experiences are crucial for further therapy sessions. If he decides at the end of the session that it wasn't too terrible, but actually quite fun, next time, he will be happy to participate in the treatment. Even the owners, who had previously believed that their dog was either afraid of water, or was very stubborn, are fully convinced about the benefits of the training after the third or fourth time. Dogs usually notice very quickly what does them good, or can quickly become convinced."

It is fascinating to watch a dog on the underwater treadmill, which allows them to work on many things at once. Their gait is often exaggerated, new muscle groups are engaged, extra balance is required on the treadmill, they are better able to bear weight on their limbs, and the treadmill aids both neuromuscular re-education and agility. The viscosity of the water creates resistance for the muscles, resulting in a great strength and endurance workout.

Summary

So now, you and your dog have a therapist, a treatment programme, and a few weeks of rehab ahead of you. The therapist will go through all the exercises and give you 'homework' to do with your dog. In addition to this, the next part of the book shows you a great collection of do-it-yourself toning exercises that I used for Argos' rehab. These exercises are not intended as a replacement for the highly effective aqua-jogging, but they will show you practical ways you can play with your dog and train him appropriately without wearing him out.

Depending on where you walk (forest or park?), you can choose which exercises you want to do and give them a go. Some exercises are suitable for inside, or for the garden; experiment and see which ones your dog likes best. And remember: your dog doesn't mind whether you call it training, physiotherapy, rehabilitation or physio-homework. He's happy whenever you do something together that's varied and enjoyable.

Chapter 10

Exercise to rebuild muscle

After the operation, clarify with the vet when you can begin various exercises with your dog. Although we have included some instructions on how often you should do these exercises, this is intended as a guide only. Every operation is different, and the healing process varies from dog to dog, so it's essential you get advice from your vet.

Of course, your dog will not appreciate that these exercises are for his benefit, so you will have to make them attractive to him.

Encourage him with catchphrases such as "activity time!" which tell him that you will be playing new and exciting games! For most dogs, there can never be enough variety; all new opportunities are welcome and, fortunately, old games, after a few days' break, will seem new again.

Your dog will almost always work for the prospect of a reward, be this praise, attention, or food.

For Basco, who loves games, that can mean fetching a ball or playing a game with a Kong®. Aisha is primarily motivated by treats, so search games involving small treats are ideal for her. She enthusiastically searches for the food dummy and is ecstatic when she is given her treat from it. Vary the accessories and games that you use to keep your dog's interest; two or three are all you need if you swap and change them each time. If the squeaky duck disappears for a few days, and then reappears, it is much more interesting than having it to play with for five consecutive days, morning, noon, and night.

Next, you will need some little food treats, but instead of chew sticks and dog chocolate, use some of your dog's daily food allocation. Aisha and Basco are quite happy to receive dry food as treats from their daily food ration, and this will prevent weight gain.

The food dummy can be used with and without food. It is a durable, sealable bag – you could even make one yourself! (See *Useful accessories*.)

Fill the food dummy with part of the daily food ration – which will vary according to the number of walks/workouts your dog does. If you are taking three walks, each time you go, fill the dummy with a little less than a third of the daily ration; for four walks, a little less than a quarter. Whatever is left over from this portion can be kept in reserve for special treats, until after the last walk in the evening, or when doing additional exercises at home.

It may seem pedantic to divide up the food in this way, but it will help to ensure that you don't overfeed your dog. Aisha has a huge appetite, Basco drools and stares at the food dummy – nearly all dogs are very greedy, and you always have reason to reward your much-loved pet, but your dog shouldn't become overweight as a result.

Every gram of excess food puts unnecessary strain on the body, joints and ligaments. They become overburdened, as does the circulation, thereby reducing your dog's desire to do something new with you and train his muscles.

Being slightly underweight is healthier than being even slightly overweight. Take advantage of every vet visit to get your dog weighed, and ask the vet whether he thinks your dog's weight is okay. If necessary, he will be able to advise you how best to help your dog lose weight.

Ask your vet how long after the operation you can begin with simple exercises, how long your walks should ideally be, and how to extend the time gradually.

The best way to control your dog's movements is to use a chest harness (see *Useful accessories*) to which you can attach a short lead. At the start of the walk, in the warm-up phase, allow him to determine his walking speed – but watch for signs of pain or discomfort. When outside, your dog must first re-familiarise himself with walking. His weight may be unevenly distributed, or he may feel unsure, confused by the 'foreign' feeling of walking again.

When venturing out for the first time, walking on his 'new' leg will feel really odd to your dog. He will probably be unsure or unsteady at first, so be kind and patient with him. Gently encourage and reassure him by talking to him whilst you are walking.

TIP

Always have the food dummy to hand whenever you are on the move. Allow your dog to carry it – perhaps to the nearest bank or the next woodpile, or wherever you want to do an exercise during the walk. Have a toy (for example, the Kong®) in your pocket, or put a small toy in the food dummy, which your dog can search for, fetch, and finally play with the contents.

There should always be a reward immediately after a new exercise is performed. Then, to keep it unpredictable, reward him after two or three repetitions, and later, after four, two, five, one, three repetitions. Always give a kind word of praise when he has done something correctly.

If he has done something wrong, ignore his mistake; a cry of "No!" is too negative, and your dog will feel disappointed or discouraged. Simply start again.

Immie has two types of harness. She wears this one when out walking ...

... and this one is her safety harness when travelling in a car. A special strap fits around the seatbelt and then attaches to the D-ring (which means, of course, that the harness can also be used when walking).

Hindquarter weight-bearing

Lead your dog to a bench (or a low wall), and place a morsel of food or the food dummy on the edge, so that he can place his front paws on the bench and reach up to the item.

With a small dog use (according to his size) a tree stump, a tree trunk, or possibly a large rock.

If you know that your dog likes to jump up, then hold his harness, and say "stay" while you hold the food just above his nose on the bench, so that his front paws remain on the edge. Now, raise the treat higher so that your dog has to stretch – and possibly even lean backward slightly – to reach his treat.

As your dog stretches for his treat, look at the toes of his hind legs, which should be spread a little as he weight-bears on them. If the toes of only one foot are spread, or neither are, it could be that something hurts, or maybe he's afraid that it might hurt to do this.

Observe his toes on a regular basis when he does this exercise – you should find after some time that the toes are spread, which is a sign that your dog's legs are getting stronger. You *both* deserve a treat for this!

Argos loves this exercise so much that he will carry the food dummy to every bench he sees, even in the dark, drop the dummy, put his front paws on the bench, and wait for his treat, as if to say "again, please!"

TIP

Ask your dog to carry the dummy (with a treat inside) the last 5 or 10 metres towards a bench (or low wall), and drop it on the bench. Have him stand with his front paws on the bench, and hold the dummy above his head, and he should bend quite far forward and then backward (which helps with co-ordination). Open the dummy and hold the treat above his head, so that he leans forward and back again. Finally, give him the treat.

Fetching, dropping, bending forward and back, eating, is an entire exercise sequence that you can vary so that it never becomes boring.

A low tree stump is suitable for helping to improve weight-bearing on the hindquarters. Try this exercise for a few seconds at a time initially, and slowly increase the duration.

In the forest where we walk, there are tree trunks next to the path. Aisha is encouraged to stand on the tree trunks: front paws on the trunk, back legs on the ground, stretched out as far as she can.

You may sometimes find stacks of logs by the side of the path. Basco walks along one of the lower logs and likes to stretch up to investigate the next log. Encourage your dog to stretch a little higher and for a little longer by offering a treat. Don't actually allow him to climb the log pile, of course, he should simply stretch up. Encourage him to do this by holding treats just within his reach.

If the benches start to become boring, find a tree with very rough bark, such as a pine tree. Estimate how far your dog can reach up the tree when standing on his hind legs, and lodge a treat in the bark

Log piles are ideal for training canine muscles.

How high can your dog reach up the tree when he stands on his hind legs? This game is ideal for strengthening his hindquarters.

tree, he will investigate by standing upright against the tree and leaning to the right and left, because something smells tasty – which is the perfect way to strengthen his legs. Put a morsel of food on a branch, wedge the dummy in the bark, or hide a few crumbs in a hole in the tree. You can increase the length of time he stands upright by putting something on the bark which he can lick off – perhaps a piece of banana or a dollop of wet food.

If your dog loves scentwork, use a piece of kibble (dry meat or fish) in the tree, and drag the food dummy along a track while your dog is otherwise occupied. Lead your dog to the start of the track, say "find" and let him follow the track and reach up the tree as high as he can until he finds what he's looking for.

There are other ways that trees can be used as exercise aids. Try this one for which you will need a toy with a rope attached to it. Hang the rope on a branch, and allow the ball to temptingly dangle from the tree. Check the height – if your dog can reach up to the ball while standing with his hind legs slightly bent, it is in the right place. Ask him to get the ball; make it seem very exciting. If he retrieves the ball, give him lots of praise, but if he's been trying for a while without success, ask him to sit and then

at this level. He then has to search for his treat. If he has to scratch or gnaw on the bark to get the treat between his teeth, that's great, as he will give his legs a good workout – and you should both find this exercise quite enjoyable. Getting him to stand up on his hind legs may not be easy, but it's worth it, for the sake of his muscles.

If he gets the idea that some trees have treats in the bark, when you lead him towards the next praise him for being so good! I can always tell if Argos has had enough because he sits down and looks at me, as if to say, "Well, I can't get it – you try!" So end the game before your dog becomes frustrated or bored. If he has happy memories, he will be pleased to play the game again.

Caution!

I had the bright idea of extending the rope using the lead, leaving the ball dangling over a ladder rung, while I held the lead from behind the ladder. So far, so good – my dog pulled at the ball, I pulled at the lead, and he pulled and growled and fought. Great. Until he suddenly let go of the ball and I fell on the ground with a thud! If you play tug of war with your dog in this way, please be aware of what might happen!

Hindquarter flexibility

With your dog standing face-to-face in front of you, attract and hold his attention with a toy held above his nose. Move it gradually forward over his head toward his hindquarters, and he should bend his hindlegs, as if to sit down – this is precisely the moment where you praise him.

This half-squat position is hard to teach and hard to learn, because dogs prefer to sit, which they are more accustomed to. If it doesn't work, then it doesn't matter; there

Playing tug of war with front legs off the ground.

are plenty of other exercises you can do instead. But if it works, get your dog to hold this position for 2 or 3 seconds – great!

Have your dog stand facing a bench, and throw some treats underneath it. To look under the bench to locate the treats your dog will have to bend down on his front legs, and then bend his hind legs, too, to get to the treats. So that he doesn't just lie down, get down with

TIP

If you have already used a clicker, you will know that you can 'tell' your dog exactly what you want him to do with it. If not, see *Useful accessories* for more information.

The half-sitting position – virtually a squat – is a great workout for the hindquarters, but initially difficult for the dog to learn.

Crawling under a bench requires good co-ordination and is suitable only for experienced patients. If your dog is reluctant to do this, please do not try and force him.

him and hold his stomach off the floor with your hands underneath him. After a few repetitions, he will understand what's required.

Using a treat, encourage your dog to creep halfway under the bench, then tempt him with a toy, which he should try to grab and pull out from under the bench. Once he has the toy in his mouth,

gradually loosen your grip on it. Just before he's about to give up, let him have the 'prize' and praise him.

In this exercise, Argos tries to fool me by lying down on the ground and then sliding under the bench. I don't give him his treat – instead, I push my toes under his tummy so that he can't get comfortable and has to hold himself up with his leg muscles.

Through the tunnel

Tunnels (fabric tubes) are ideal for your dog to run through. It's better if he can't run with his head held up, but has to crawl instead, which is a great exercise for his leg muscles (see pages 24 & 25).

If you don't have a tunnel, ask your dog to crawl under a bench or under the sofa, chair or, better still, several chairs in a row. Or you could form a tunnel with your legs and get him to go through it (sit on the floor and put your legs or feet up on the sofa, or sit on the edge of a chair with your knees bent to make a tunnel).

If you have children, have them stand one behind the other and create a tunnel with their legs. Your dog can run through the tunnel and your children can gradually make that tunnel smaller and smaller so he has to crouch to get through.

Balance exercises

After surgery, your dog needs to strengthen his muscles and train his reflexes. Balance exercises will improve his proprioperception (the sense of spatial awareness and limb positioning).

With regard to your own body, you will automatically know, for example, whether your leg is straight or bent; you're not even aware of thinking about this as it is done subconsciously. Without proprioperception, body control and co-ordination would not be possible: not for gymnasts, ordinary people, or other animals.

After the operation, your dog's proprioperception will be a little confused. With Argos, this was very noticeable: after the operation, he couldn't balance on a log like he could before, and I had to slowly help him learn proprioperception all over again. At first, I got him to run on thick branches with very rough bark (so that he wouldn't slip), while I stayed close by, ready to catch him if he fell.

Safety first, as always! As long as Argos is still unsteady on his legs, I ensure he doesn't slip and become frightened, anxious, insecure or injured.

Aisha went to dog-dancing classes for a while – to give you an idea of what this entails, the following are examples of a few basic exercises:

• spins, ie small circles to the right or left
• slalom (running between her owner's legs)
• a figure of eight (running between her owner's legs from right to left and vice versa), or
• running backwards in a circle around her owner

Practising balance on a log.

Teach your dog how to balance with the help of a car ramp or a ladder.

TIP

This simple exercise will help your dog feel at home again in his own body.

Ask your dog to sit and ask for first one paw, then the other. Keep alternating the paws and this will help him regain his balance and build his confidence.

Your dog will understand best what you want from him if you teach him using treats and lots of patience. For example, going back to the log piles, if they aren't too high, just let your dog investigate them. Hide a little treat in-between the logs so that your dog has to balance on his hind legs to sniff it out. Wobbling back and forth along the logs and sniffing for his treat will improve his strength and balance. Stay close to him in case he panics or slips, though.

Is your dog small to medium in size? If you have a mini trampoline, see if he will jump on it. Let him climb carefully onto the trampoline (for a very small dog, use a board as a ramp). It may take him a while to adjust to the different surface, which will seem very strange at first. Give him time. He will eventually get used to it and maybe begin to enjoy this unusual exercise!

Can you encourage him to stay on the trampoline for a little while? Fantastic. Make sure you give your four-legged friend lots of praise!

Can you get him to walk a few steps on the trampoline? With food

Right: On a woodpile, different stresses are applied to each leg, which will help him to relearn proprioperception.

in your hand to tempt him, and with patience and plenty of praise, see if he will walk round and round the trampoline, clockwise and vice versa – great! Really praise him if he is willing to do this.

After Argos had become familiar with standing on the trampoline, I carefully depressed the surface of the trampoline, then released it so that Argos was bouncing a little. I made sure that the soles of his feet stayed in contact with the surface while he used his balance to stay upright. The following isometric exercises can be done either on or off the trampoline.

Isometric exercise
Isometric exercise, or isometrics, are a type of strength training in which the joint angle and muscle length do not change during contraction (compared to concentric or eccentric contractions, called dynamic/isotonic movements). Isometrics are done in static

Do you maybe have a mini trampoline your dog can use?

positions, rather than being dynamic through a range of motion. The joint and muscle are either worked against an immovable force (overcoming isometric), or are held in a static position while opposed by resistance (yielding isometric).

Examples of this are pushing or pulling an immovable object like a wall, or a bar anchored to the floor. Although research shows that isometric exercise increases muscle strength significantly, it doesn't change the length of the muscles. Today, it is primarily used for rehabilitation purposes. The exercises strengthen the muscles and train the reflexes: you can use these on a healthy dog to help prevent hip dysplasia, for example, and also with a dog who has had surgery, to strengthen his muscles afterward.

Here's an example of such an exercise:

With Argos standing in front of me, I place one hand flat on his right thigh, and the other on his right shoulder. Now, with both hands I press gently against his body. You only need enough force so that your dog resists the pressure to prevent himself from falling over. Your dog stays on the same spot, using his muscles to resist the pressure of your hands. Holding this position, slowly count to five, gradually let the pressure subside and praise your dog. Carry out this exercise three times in a row, several times a day, swap hand positions; that's it! This is a good exercise to add to your rehab programme: doing this on the trampoline makes it slightly more challenging as your dog has to retain his balance as well as resist the pressure.

Standing on three legs

Make sure your dog is standing on firm ground, then lift one leg at a time and hold for a few seconds (ask your vet or physiotherapist to show you how to do this safely). You can do the same on the trampoline, but please ensure that your dog isn't anxious and is able to keep his balance.

Slow steps

Walk your dog at a very slow pace, which will ensure that he uses all of his legs and can't leave out his weakest leg. The best time to do this is after he has done a few other exercises or been for a walk. Put your dog on a short lead and walk slowly enough that he can't break into a trot.

The smaller the dog, the more difficult this will be as he will be more likely to break into a trot.

Left & opposite: Soft surfaces are best for your dog to exercise on.

TIP

Allow your dog to run on soft surfaces as often as possible: in meadows and in the forest, on soil, sand, and through snow. If you are near the sea or have a stream nearby, this is also good, but only if the surgical wound has completely healed.

If you have access to a schooling ring (inside or outside), take him there as the floor is usually the ideal surface (wood shavings, sand, or a sand and fabric blend, wood chips, etc) on which to walk your dog.

Walking uphill

At first, only practice walking uphill sparingly, and don't increase this until you have achieved some progress with your dog's training. As a precaution, ask your physiotherapist or consult with your veterinarian for advice on this.

Good places to practice uphill walking are small – or even large – banks, for example. Keep your dog on a short lead, in case he tries to charge up the hill. Also, walking up a ramp but be sure to stay right beside him in case he should seem unsteady.

Walking uphill requires more muscle power, so you may prefer to use a faster gait. You'll see that your dog walks uphill with a high, curved back – this powerful force exerted by his muscles is a great workout for him. Vary the exercise by going straight up a few times, then diagonally or zigzag. This not only strengthens the muscles of your dog, it will 'feed' his nerves with information about proprioperception.

Walking up a ramp.

Also vary the pace – and the place you do this – but always ensure that your dog remains in step with you.

Walking backwards

If your dog is unfamiliar with this (as I expect most dogs will be), begin by walking alongside a fence. After a few metres, stop walking and hold a morsel of food above his nose and slightly over his back. His first reaction may be to sit down, so prevent him from doing this by holding his harness up a little. He should then take a backward step in order to reach his treat. Give him his treat straightaway and lots of praise. Repeat this again, until

he takes a few steps backward. It doesn't matter what word you use for this; you can simply say 'back' or 'backward,' for example.

Climbing steps

This is a good exercise to strengthen the hindquarters. Begin slowly with low steps and, as your dog's strength increases, progress to steeper steps. Put him on a short lead and harness so you can ensure that he uses controlled movements only.

Caution!

Check that your dog lifts each of

Keep the lead quite short for a controlled ascent and descent.

his paws properly, step by step, and doesn't leave out a particular leg, or jump or hop.

Sit-and-stand exercises

Whilst on a walk, ask your dog to sit and then get up again quickly, which will exercise and train the hip and knee extensor muscles. While he is sitting, check whether both hindlegs are in the same position, or if he is stretching one leg to the side. So, if he stretches his left leg to the side, walk him with the left side of his body close to a wall, then have him sit close to the wall so that he can't stretch out the left leg. Praise him when he's sitting correctly, even if he does so only because the wall prevents him from strtetching out the leg.

Tell your physiotherapist if he stretches one particular leg to the side, because she will know what to do in order to avoid or lessen any permanent deformity.

Lifting the paws

When your dog walks, he should lift his feet properly and not slouch (just like us!). To improve his walking, place some logs in his path and walk him over the logs on a short lead. The logs are the correct distance apart when your dog has to lift his feet high to step over them. If the position of the logs matches his stride, repeat this several times with him. Make sure that he doesn't bounce or jump over them, though.

If you have a very small dog, you could lay out a ladder for him to climb over. Walk your dog over the rungs of the ladder but don't allow him to jump. This is not a race, more a balancing act.

Caution!

The best type of toy to fill with treats is one made of rubber, which doesn't split when the dog bites it.

Opposite, top: Obstacle course at a walking pace (with treats as a reward).

Opposite, bottom: Obstacle course at a trot (with the ball as a reward).

TIP

Really make the most of this next exercise; not only will it help your four-legged friend convalesce, it will also remind him of his manners!

If I want something, I say please. When my dog wants something, he should say please as well. Before you take your dog out for a walk, or give him his dinner or a cuddle, ask him to sit. This way, he is learning to say 'please' and strengthening his muscles at the same time!

TIP

How about tying a ribbon or a cat's collar with a bell around the leg which has been operated on? Some dogs will find this irritating, and will lift their leg higher as they walk. Perfect! Take advantage of this response – fill a Kong® with some treats so that he's so engrossed in trying to get at these that he leaves the ribbon alone.

It's also very important to use a big enough toy so your dog can't swallow it. And the treats should be the ideal size, too; big enough so that only one falls out each time, not several all at once.

Use a portion of his usual food allowance, so that he has to work for it.

Argos loves his evening ritual: after the last walk, I fill his snack ball with the last of his daily rations. Before I do this, I put a resistance band (see *Useful accessories*) on the foot of the operated leg, and while he is eating his last meal, he does an extra quarter of an hour's training.

Swimming

Swimming is a perfect exercise for your dog – but only once the surgical wound has healed completely. Take advantage of lakes and streams, or wherever else dogs are allowed to swim in your area.

Do you have a paddling pool? If the weather is warm enough, and your dog is small enought to swim in it, fill it up and let him get in. Pull a toy on a string through the water, so your dog swims to catch

the 'prey.' Trying to catch the toy will increase his motivation to swim.

Caution!
The floor surface of paddling pools is often too slippery for dogs. If your dog should slip, not only could this injure him, it could deter him from this exercise. Lay a non-slip mat on the floor of the paddling pool.

Argos likes to swim, but we have only a very small pond nearby. I throw a dog chew (it floats!) into the water as far as I can. He swims out, grabs it, and then I throw another one in a different direction, and so on. On the way to getting the treat, he splashes about and gets plenty of exercise.

Caution!
Fetching or eating while swimming could cause your dog to swallow large amounts of water. Be aware that, over the next few hours, he may need to urinate more often than usual!

Does your dog love fetching, like Aisha? Then an unsinkable frisbee is perfect for him! (see *Useful accessories*). If you buy a disc, keep in mind when choosing the colour that dogs can see a light colour on a dark background, and

Opposite: Swimming is an ideal way to train your dog without putting stress on his joints.

can't distinguish between red and green. Yellow and blue are the best colours to choose.

Caution!
Throwing a stick, ball or frisbee on land is absolutely taboo, because they encourage the dog to jump up, and make sudden stops and changes of direction. If you use balls in the water then avoid tennis balls – they float and are easily visible, but are coated with a material that can be harmful to canine teeth. Make sure that the ball is sufficiently large that he doesn't get it stuck in his throat, which could cause him to choke. For a German Shepherd, a ball the size of a tennis ball could be disastrous.

Strengthening the spine and front legs
Because your dog may have developed poor posture due to his injury, you will also need to take care to strengthen the remaining three paws and his back. Specifically targeted exercises can strengthen the spine and front legs, as shown in the picture overleaf. A physio-ball can be used to stretch the spine, strengthen it, and also train the front and/or back legs. By doing tight turns, your dog will become more co-ordinated and exercise his spine at the same time.

In the foregoing text we have

tried our best to describe the exercises used for muscle building in clear and concise terms. The accompanying photos will enable you to see the exercises in practice, and also help you to remember each move. You can vary an exercise according to your environment.

Doing the same thing all the time isn't a good idea – and this applies to dogs just as much as humans: Argos, Aisha, Basco, and Chico would agree with this. Because you will need to keep doing the same *type* of exercise, vary the actual routines day to day. Sometimes, just a different kind of reward

Training on the physio-ball.

(apple slices instead of dry food), or another toy is enough of a variation. Argos loves nothing better than a toy which grunts or squeaks. I've always got a squeaky toy in my pocket in case of an emergency. If he seems fed up or is refusing to do any more, he can chew on the toy for a while as a reward. This will help to keep him motivated and interested in what he is doing.

Life, and certainly life with a dog, is too precious to waste doing something joyless or tedious. Find out what makes your dog tick and use this to motivate him during training. Having fun with your dog during physiotherapy and rehabilitation is most definitely possible.

Hopefully, this book will have shown you how you can make training fun and varied. We wish you every success and happiness, and your dog all the best for his recovery.

Slow, tight turns are good for weight-bearing on the injured limb, and also improve coordination.

Appendices

Useful accessories

Aquatherapy

You can obtain information about canine aquatherapy or hydrotherapy in the UK from the CHA (Canine Hydrotherapy Association): www. canine-hydrotherapy.org. Under the CHA Member section, you will find contact details of all approved UK canine hydrotherapy centres.

Visit www.iaamb.org for information about US-based organisations.

Bandages

This website is an online chemist for dogs! It sells almost anything you can think of, including bandages and dog socks to protect the bandages: www.bestpetpharmacy. co.uk.

Child gate/stair gate

This particular child gate comes very highly recommended. It has won several awards for its safety features and is a reasonable price, too: www.argos.co.uk/static/Product/partNumber/3763461.htm.

Classes

To find dog training classes in the UK, simply visit this website and type in your postcode or county: www.kennel-corner.co.uk.

To find out more about dog-dancing in the UK, visit this website for information and videos: www.dancingdogs.co.uk.

Clicker training

A conditioning method of dog training, a clicker, or small mechanical noisemaker, is used as a marker for behaviour. The method uses positive reinforcement because it is reward based. The clicker is used to teach a new behaviour, to enable the dog to rapidly identify that a particular behaviour is required in order to hear the noise of the clicker, and therefore receive the reward. See the following websites for more information:

www.fun4fido.co.uk has an excellent, in-depth guide to clicker training.

www.clickertrainusa.com

www.seapets.co.uk sells clickers which come with a free guide and training tips.

There are also plenty of great clicker training videos on YouTube: www.youtube.com.

Cold/hot packs

You can buy these from any chemist and most supermarkets sell them, too. The best ones to look for are the gel packs, as these are flexible and can be wrapped around the leg, for example.

Dog ramp

Dog ramps can be very costly, but this one is a good price and comes highly recommended; visit www.bargainbrands.co.uk/dog_ramp.html.

This company makes a foldable, lightweight wooden ramp which has been recommended on several pet forums; visit www.overthetop.co.uk/shop/Tri_Fold_Car_Pet_Ramp.html.

Dog wheelchairs

www.doggon-uk.com is an excellent website which sells many types of products to help your dog stay mobile. The many customer testimonials and helpful photographs will enable you to choose the right product.

You can also buy these products second-hand on ebay: www.ebay.co.uk.

Food dummy/training dummy

This website offers four different varieties of training dummies, available in a choice of colours and with a sealed interior so that it will float in water: www.countrykeeper.net.

Frisbee

Available from any good toy shop. Also see below.

Intelligence games for dogs/Kongs®

www.zooplus.co.uk/shop/dogs/dog_toys_dog_training/intelligence_games is a great source for all things dog-related, including accessories and intelligence games to test how clever your dog is!

Harness

www.doggiesolutions.co.uk has a large variety of harnesses to suit every dog.

Physiotherapy

www.acpat.org is the association of chartered physiotherapists in animal therapy in the UK. This website will help you find a reputable canine physiotherapist in your area.

Your vet should also be able to recommend a physiotherapist.

For US organisations/therapists, see www.caninerehabinstitute.com.

Further reading

Swim to recovery: canine hydrotherapy healing – Gentle Dog Care by Emily Wong. Published by Hubble and Hattie. ISBN: 978-1-845843-41-0.

The complete dog massage manual – Gentle Dog Care by Julia Robertson. Published by Hubble & Hattie. ISBN 978-1-845843-22-9.

Dog Relax: relaxed dogs, relaxed owners by Sabina Pilguj. Published by Hubble and Hattie. ISBN: 978-1-845843-33-5.

Exercising your puppy: a gentle and natural approach – Gentle Dog Care by Julia Robertson and Elisabeth Pope. Published by Hubble and Hattie. ISBN: 978-1-845843-57-1.

Living with an older dog – Gentle Dog Care by David Alderton and Derek Hall. Published by Hubble and Hattie. ISBN: 978-1-845843-35-0.

Animal physiotherapy: assessment, treatment and rehabilitation of animals by Catherine McGowan and Narelle Stubbs. Published by Wiley-Blackwell. ISBN: 978-1-405131-95-7.

My dog has hip dysplasia – but lives life to the full! by Kirsten Häusler and Barbara Friedrich. Published by Hubble and Hattie. ISBN: 978-1-845843-82-3.

My dog is blind – but lives life to the full! by Nicole Horsky. Published by Hubble and Hattie. ISBN: 978-1-845842-91-8.

My dog is deaf – but lives life to the full! by Jennifer Willms. Published by Hubble and Hattie. ISBN: 97-1-845843-81-6.

My dog has arthritis – but lives life to the full! by Gill Carrick. Published by Hubble and Hattie. ISBN: 978-1-845844-18-9.

Emergency first aid for dogs – at home and away by Martn Bucksh. Published by Hubble and Hattie. ISBN: 978-1-845843-86-1.

Index

Gentle Dog Care

128 pages • 100 colour photos
• 20.5x20.5cm • ISBN: 978-1-845843-22-9
• £12.99*

the complete dog massage manual

Julia Robertson

Hubble&Hattie

Gentle Dog Care

swim to recovery
canine hydrotherapy healing

Hubble&Hattie

Emily Wong

Gentle Dog Care

128 pages • 130+ colour photos
• 20.5x20.5cm • ISBN: 978-1-845843-41-0
• £12.99*

For more info on Hubble and Hattie books, visit our website at
www.hubbleandhattie.com
email info@hubbleandhattie.com • tel 44 (0)1305 260068 • *prices subject
to change • p&p extra

78

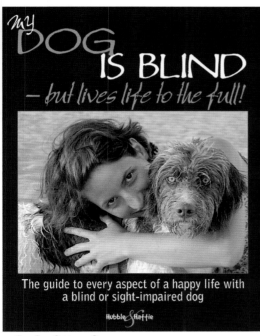

my DOG IS BLIND
– but lives life to the full!

The guide to every aspect of a happy life with a blind or sight-impaired dog

Hubble & Hattie

978-1-845842-91-8

my DOG IS DEAF
± but lives life to the full!

The guide to every aspect of a happy life with a deaf or hard-of-hearing dog

Hubble & Hattie

978-1-845843-81-6

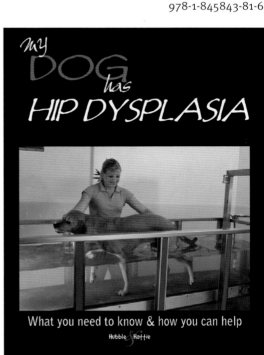

my DOG has HIP DYSPLASIA

What you need to know & how you can help

Hubble & Hattie

978-1-845843-82-3

All 80 pages • c50 colour photos • 22x17cm • £9.99* each

Important notes